Process Development Division 2024

Held at the 2024 AIChE Spring Meeting and 20th Global Congress on Process Safety

New Orleans, Louisiana, USA
24 - 28 March 2024

ISBN: 979-8-3313-1706-5

Printed from e-media with permission by:

Curran Associates, Inc.
57 Morehouse Lane
Red Hook, NY 12571

Some format issues inherent in the e-media version may also appear in this print version.

Printed with permission by Curran Associates, Inc. (2025)

For permission requests, please contact AIChE
at the address below.

AIChE
120 Wall Street, FL 23
New York, NY 10005-4020

Phone: (800) 242-4363
Fax: (203) 775-5177

www.aiche.org

Additional copies of this publication are available from:

Curran Associates, Inc.
57 Morehouse Lane
Red Hook, NY 12571 USA
Phone: 845-758-0400
Fax: 845-758-2633
Email: curran@proceedings.com
Web: www.proceedings.com

TABLE OF CONTENTS

ADVANCED CARBON REDUCTION TECHNOLOGIES

ADVANCES IN PROCESS DESIGN AND DEVELOPMENT

ADVANCES IN PROCESS DESIGN AND DEVELOPMENT II

COST-EFFECTIVE COMMERCIALIZATION OF NEW PROCESSES OR PRODUCTS

MODELING TOOLS AND TECHNIQUES FOR PROCESS RANDD I

MODELING TOOLS AND TECHNIQUES FOR PROCESS RANDD II

POSTER SESSION: PROCESS DEVELOPMENT DIVISION

Ethylene Production and Decarbonization Strategy Using Oxidative Coupling of Methane

Balaji Krishnakumar, Himant Srivastava, Edward Katende*

Lummus Technology LLC
**Balaji.krishnakumar@lummustech.com*

Prepared for Presentation at the 2024 AIChE Spring National Meeting
New Orleans, LA, March 2024

Abstract:

Oxidative Coupling of Methane (OCM) upgrades low-value methane-rich feedstock into high value petrochemicals. Latest improvements in Lummus' Gemini® OCM technology bring it closer to commercial deployment on a cost-competitive basis. It also offers an avenue for reduction in CO_2 emissions. OCM can be deployed in a stand-alone mode using natural gas and gas liquids as feedstock; integrated with a Gas Plant; integrated with a steam cracker; or utilize CO_2 as a feedstock. Different process schemes can be employed for OCM when external H_2 source, particularly from a green source such as electrolysis is available, CO_2 itself can be used as a feedstock to produce green ethylene using OCM. Some of the salient features of this technology and the different flow schemes will be discussed in this paper.

Introduction

Methane is an extremely stable molecule with strong C-H bonds and is not susceptible to chemical activation. As a result, the splitting of the C-H bonds requires high temperatures (>900°C), and/or the use of oxidation agents to facilitate conversion. Siluria® OCM (oxidative coupling of methane) Technology has been previously presented [1, 2, 3, 4] and is now owned by Lummus Technology. Since acquisition, Lummus has resolved several technical hurdles pertaining to the OCM reactor bringing the technology to commercial readiness.

Lummus Technology currently offers two branded main olefin-producing technologies via the oxidative coupling of methane (OCM); Gemini™ Process (Ethylene production via OCM process) and the Orion™ Process (Propylene production via OCM). The Orion™ process couples a dimer/OCT unit to the end of the Gemini™ process train to provide product flexibility. The input stream for both processes is low value methane-containing streams such as, natural gas, off-gas from various petrochemical facilities, propane de-hydrogenation units, and refineries.

Lummus' OCM offers a direct route to upgrading CH_4-rich fuel streams to ethylene unlike multi-step routes through synthesis gas, a mixture of H_2 and CO, to methanol/olefins or directly to olefins. In the OCM process, methane/natural gas and oxygen react over a catalyst to form ethane and ethylene (C2) and small amounts of higher hydrocarbons (C3+ products).

OCM Reactor

At the core of the Gemini™ process design is the OCM adiabatic reactor, shown in Fig-1, which consists of two main sections: the OCM Reaction Section, which contains the OCM catalyst bed, followed by an adiabatic chamber designed to perform Post Bed Cracking (PBC) of ethane and/or propane. The reaction vessel is refractory-lined.

The feed to the OCM reactor, primarily consisting of methane, is mixed with oxygen and sent to the catalyst bed. The gas is stoichiometry is deficient in O_2 to allow partial oxidation of hydrocarbons and methane coupling. The partial oxidation generates reaction heat allowing the coupling and cracking reactions to take place. The main reaction that takes place in the catalyst bed is shown in the following reaction 1:

$2\ CH_4 + O2 \rightarrow C_2H_4 + 2H_2O + \text{heat}$ **Reaction 1**

In addition to methane oxidation, Siluria's OCM catalyst simultaneously promotes the oxy de-hydrogenation of any ethane present in the feed according to the following reaction 2:

$C_2H_4 + 1/2O_2 \rightarrow C_2H_4 + H_2O + \text{heat}$ **Reaction 2**

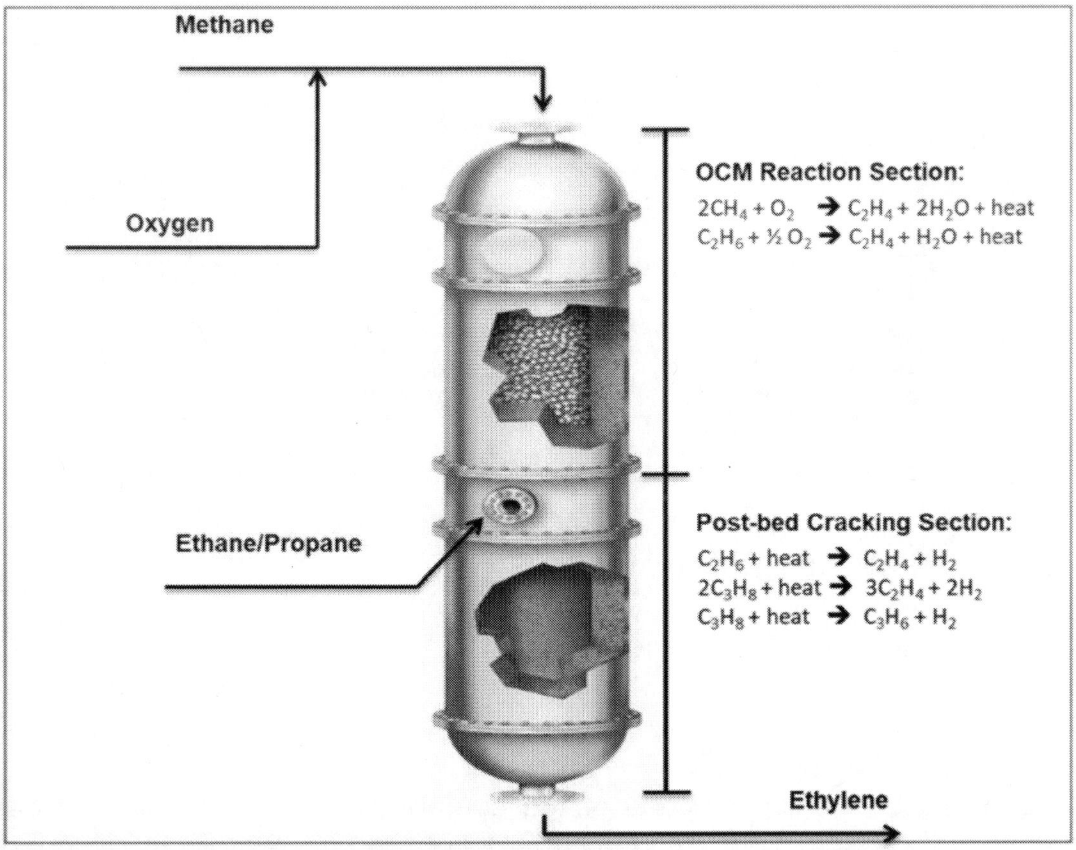

Figure 1: Schematic of OCM reactor showing the fixed bed and cracking section.

Note however, that these reactions follow a complicated reaction pathway and produce non-selective side products. Following the catalytic reaction section is the PBC (post-bed cracking) section. Ethane and/or propane can be injected into the PBC to utilize the available heat to crack the alkanes according to the following reactions 3, 4, 5.

The OCM process itself produces some net C_2H_6, which is also routed to the PBC section. When this internal strem is supplemented with external ethane and/or propane, the configuration is called C2Max. When no supplemental ethane/propane streams are available, the configuration is called C1Only.

$C_2H_6 + heat \rightarrow C_2H_4 + H_2$	**Reaction 3**
$2C_3H_8 + heat \rightarrow 3C_2H_4 + 2H_2$	**Reaction 4**
$C_3H_8 + heat \rightarrow C_3H_6 + H_2$	**Reaction 5**

OCM Process Flowsheet

A schematic of the OCM flowsheet is shown in Fig-2.

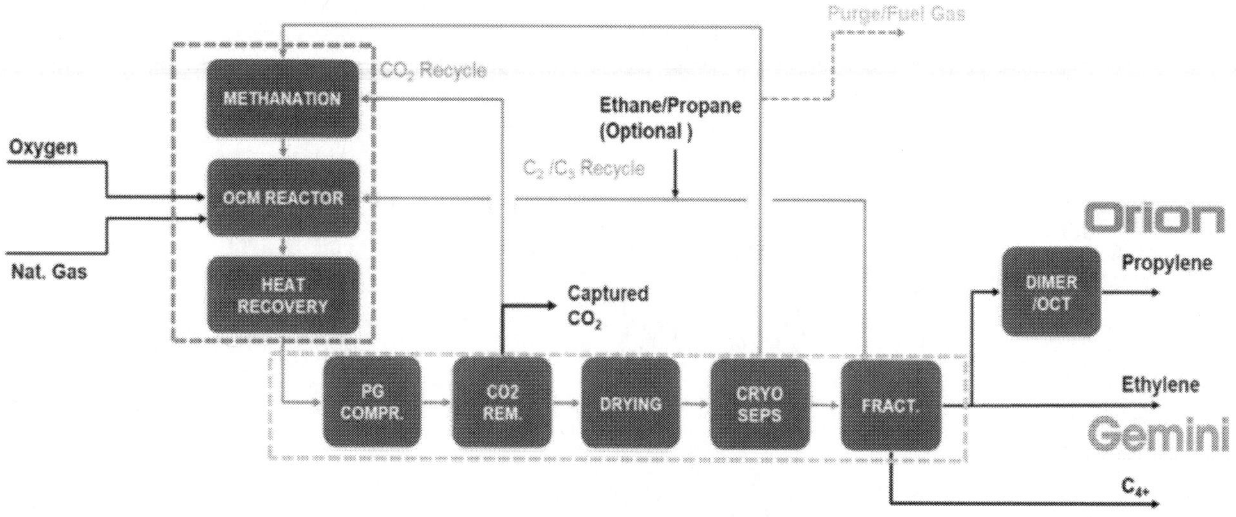

Figure 2: Lummus' OCM block flow diagram

It consists of two main sections:

1. The Hot Section which consists of the methanation, the OCM reactor and the Heat Recovery (red box section in Fig-2)
2. The Cold Section which consists of the compression, followed by acid gas removal, and cryogenic separations and fractionation (grey box section in Fig-2)

In the hot section, the recycle stream from the cold section consisting of CH_4, H_2 and CO is mixed with a portion or all of the CO_2 from the acid gas removal unit, and are fed to the methanator. The CO and CO_2 are converted to methane, to consume the available H_2. CH_4-rich fresh feed, oxygen and the methanator effluent are then mixed upstream of the catalyst bed. Post the OCM reactions on the catalyst bed, ethane and/or propane are cracked to additional olefins in the post bed cracking section as described previously. The heat from the OCM reactor is then recovered for feed pre-heating and by producing high-pressure steam.

Once the process effluent is cooled, it is compressed to about 30 barg and CO_2 is removed by acid gas removal system. The effluent is dried, chilled and sent to a demethanizer. The demethanizer overhead is recycled back to the methanator, as stated above, and the bottom stream is sent for further purification and recovery of products. Any unreacted ethane and/or propane is recycled back to the OCM reactor.

The OCM reaction does produce CO_2 by product as well as H_2. The CO_2 is recovered by the amine wash and the hydrogen stays with the unconverted methane and is recycled back to the methanator. Additional H_2 import can be used to balance the CO_2 production so that there is no CO_2 product.

Standalone OCM Configurations: C1Only, C2Max

Table 1 below, shows the material balance for a standalone OCM under two configurations C1Only and C2Max. Note that in the C2Max configuration, supplemental ethane and/or propane is provided to the post-bed cracking section to utilize the exothermic heat produced in the OCM reaction.

Table 1: Material balance for standalone OCM configurations.

All values in kTA	C1Only	C2Max
Natural gas	124	69.3
Oxygen	258	150
Ethane	-	25.3
PG Ethylene (E)	50	50
RPG Propylene (P)	9	5.8
Mixed C4	1.7	1.5
Mixed C5+	3.6	1.8
Lights + Purge	10	5.9
CO_2	116	65
Water condensate	Rest	Rest
Specific CO_2 kg/kg ethylene	**2.31**	**1.3**

For producing 50 kTA of PG ethylene, 124 kTA of natural gas feed is consumed in the C1Only configuration. When external ethane is injected into the PBC to utilize the reaction exotherm, the natural gas and O_2 consumption are reduced by more than 40 %. The reduction in CO_2 emissions is even greater. Availability of external H_2 can decrease the natural gas consumption and reduce the process CO_2 emissions to zero.

OCM using CO_2 as a feed

In a unique configuration of the OCM process, CO_2 can be used as a feedstock with accompanying H_2. A simple material balance for a 50 kTA OCM ethylene plant using CO_2/H_2 as a feedstock is shown in Fig-3. This plant would consume approximately 600 tpd of CO_2 alongwith approximately 150 tpd of H_2.

Both CO_2 and H_2 are fed to the methanation reactor to be converted to CH_4, which is sent to the OCM reactor. The OCM process also produces CO_2, which is captured and recycled back to the methanation section. Ideally, the H_2 needed for methanation must be generated via green processes like electrolysis because fossil-derived H_2 would be

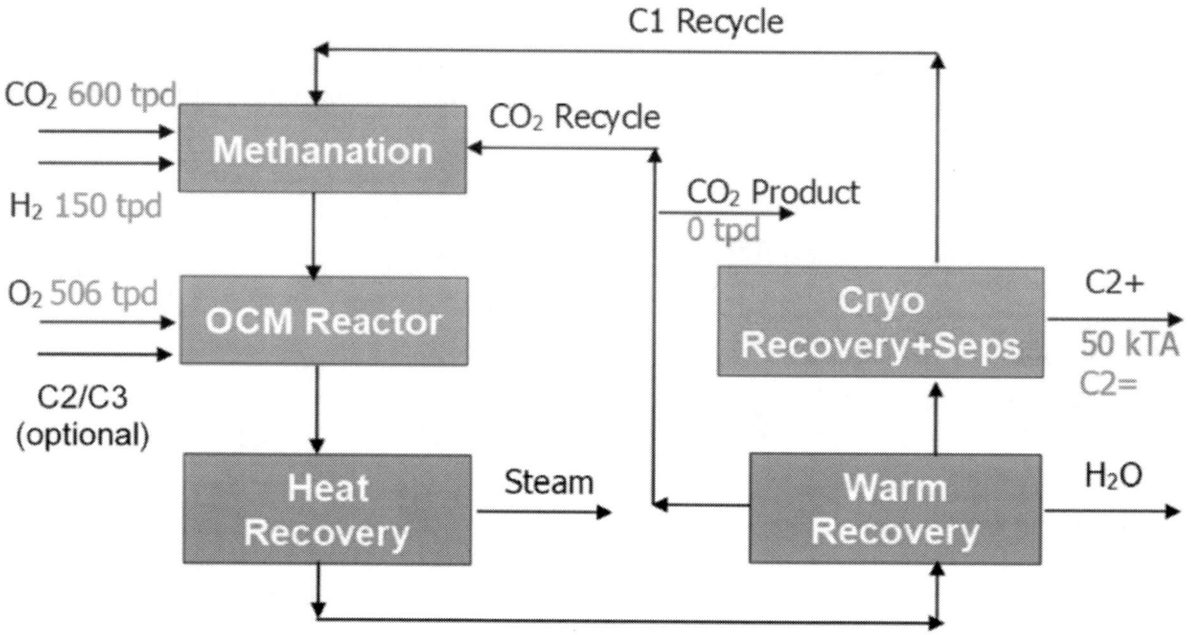

Figure 3: OCM material balance for CO_2/H_2 as feed.

counter to the objective of using CO_2 as a feedstock. Consequently, the electrolyzer would also supply the O_2 required for the process and export the excess O_2.

Since methanation reactors are commercially available, there is little technical risk in converting CO_2 to CH_4 and feeding the OCM. Such a scheme enables the use of CO_2 as a feedstock and sequester it in petrochemicals that are ultimately converted to polymers. This scheme has a negative specific CO_2 emission per kg of ethylene of <-4 since it utilizes CO_2 as a feedstock. Since the process starts with negative CO2 emissions, there is some flexibility on supplementing the feed with natural gas and/or allowing some process CO_2 emissions.

As shown in the flowsheet, the scheme is a net exporter of energy in the form of steam. The C2Max configuration can be employed by importing C2/C3 paraffin to boost the ethylene output while sacrificing steam export.

As with many cases of CO_2 valorization, the process economics for using CO_2 as a feed for OCM are challenging since it requires renewable H_2 for recycling CO_2 back into CH_4. This option may be viable with a combination of incentives for CO_2 removal, lowered cost of H_2 production, preferably from green energy-derived electrolysis, and a premium product pricing for 'green' ethylene.

OCM Integration with a Steam Cracker

Integration of the OCM process with steam cracker illustrates the feedstock flexibility of the OCM process. Steam cracker vent streams containing light end gases (the demethanizer overhead) are typically sent for fuel. These can be recovered and converted to ethylene to increase the overall ethylene production. The existing ethane and propane recycled to the cracker furnace can be directed to the Post Bed Cracking (PBC) section of the OCM reactor freeing up furnace capacity for alternative feeds. This is especially beneficial to new liquid crackers that are being designed with a mix of gas and liquid furnaces. The gas furnaces can essentially be eliminated and replaced by the OCM reactor. In addition, steam from the OCM unit can also be integrated with the existing steam system for the cracking furnaces.

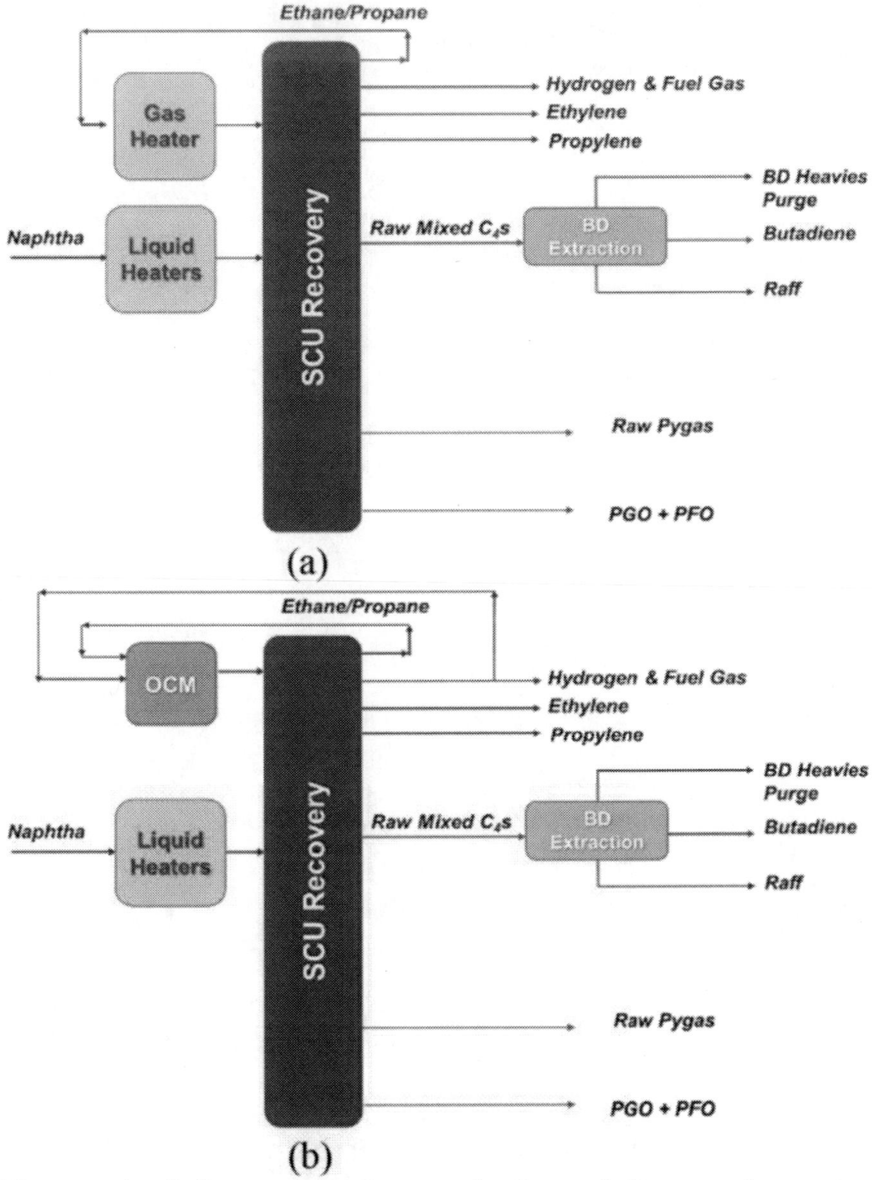

Figure 4: Schematic of a typical naphtha cracker setup (a) and OCM+cracker integration scheme (b).

OCM integration can be performed for both grassroots cracker and for existing cracker revamp. Fig-4 shows how the OCM-cracker integration would work. In a typical naphtha cracker, the liquid heaters produce significant ethane and propane that are cracked in recycle gas heaters. When integrated with OCM, the methane/H_2 offgas from the cracker is sent to the OCM reactor. The ethane/propane recycle is also routed to the OCM reactor. Hence, OCM replaces the recycle gas heaters and generates incremental ethylene by also using the offgas. Since the offgas contains H_2, it helps to reduce the process CO_2 emissions from the OCM section.

Case-I

Table 2: Material balance for OCM integration with grassroots naphtha cracker.

All values in kTA	Cracker	Cracker w/ OCM	% Change
Naphtha	4,402	4,261	-3 %
OCM DM to Cracker	-	351	-
Hydrogen + Fuel Gas to Heaters	581	532	-9%
Hydrogen + Fuel Gas to Export	140	-	
PG Ethylene (E)	1,500	1,500	-
Hydrogen + Fuel Gas to OCM	-	138	
Ethane to OCM	-	265	-
Propane to OCM	-	46	
PG Propylene (P)	644	643	-
1,3 Butadiene (BD)	227	219	-3%
Isobutene product	145	142	-2%
Raw Pygas	995	963	-3%
PGO+PFO	170	164	-3%
Total HV E+P+ BD	2,371	2,362	-0.5%
CO_2 Emission – Heater firing	**1,481**	**1,372**	
CO_2 Emission– Export Penalty	**356**	-	
CO_2 Emission – OCM	-	**140**	
Specific CO_2 kg/kg ethylene	**1.22**	**1.01**	**-21%**

Here, for example, we study a world-scale naphtha cracker producing 1500 kTA of ethylene. It also produces 644 kTA of PG propylene and smaller quantities of other byproducts. To make 1500 kTA Ethylene, the cracker consumes 4400 kTA of naphtha. It also produces C2 and C3 paraffin that are cracked in gas heaters. Overall, the steam crackers produce 721 kTA of offgas, primarily composed of CH_4 and H_2. Of this 721 kTA of offgas, 581 kTA is consumed in firing the heaters to provide the endothermic heat for cracking reactions while the excess, i.e. 140 kTA is exported. This stream is typically combusted. Considering CO_2 emissions in heater firing and CO_2 penalty on fuel gas export, specific CO_2 per kg of ethylene comes out to be 1.22.

When this 1500 kTA naphtha cracker is integrated with the OCM, as shown in Fig-4(b), the excess offgas from the cracker is routed to the OCM unit to produce additional ethylene. Ethane and Propane recycle are also routed to OCM unit, thus eliminating the need of recycle gas heater.

Now, to make 1500 kTA Ethylene in the cracker+OCM plant, the naphtha cracking requirement is reduced from 4400 to 4261 kTA, which represents a 3 % reduction in naphtha feed. The propylene production target is also met while there is an overall decrease in 0.5 % of total products, primarily because a larger reduction in fresh feed consumption. More importantly, the specific CO_2 emissions per kg of ethylene is reduced by 21 % to 1.01.

What is not shown in the table is that if additional H_2 is available to the tune of 35 kTA, all process emissions from OCM can be eliminated. This will result in a 5.7% reduction in naphtha feed requirement and a 37 % reduction in overall CO_2 emissions.

Case-II

The second case we consider is the integration of OCM with an existing cracker. We again use a 1500 kTA naphtha cracker as baseline, which will have gas heaters to crack ethane and propane recycle. When integrating with OCM, either these recycle heaters can be stopped or, be supplied with supplemental imported feed. As before, the OCM will be supplied with a portion of the cracker offgas and the ethane/propane recycle. For our evaluation, we consider importing propane to fill the existing recycle gas heater. The propane import rate is selected such that the firing duty to the heaters is kept the same as the baseline case.

With such an integration, the ethylene production increases from 1500 to 1687 kTA, a boost of 12 %. In addition, the propylene production is increased by about 9% with only a 6% increase in fresh feedstock consumption. Overall, the increase in capacity is accompanied with a 23 % reduction in specific CO_2 per kg of ethylene.

How does this excess capacity affect the existing cracker assets with regard to integration? As OCM effluent joins the cracker demethanizer bottoms, there is no impact on the cracker hot section. Thus, gasoline fractionator, quench tower, charge gas compressor, dryers and chill train and demethanizer stays same. The OCM will have its own recovery units until the demethanizer. In the C2 area, there is an approximately 15% increase in area factor, which can be accommodated with high-capacity trays in deethanizer or ethylene fractionator. We also anticipate modification in refrigeration section of existing cracker with this integration.

Table 2: Material balance for OCM integration with naphtha cracker revamp.

All values in kTA	SCU	SCU w/ OCM
Naphtha	4,402	4,402
Import Propane	-	264
OCM DM to Cracker	-	437
Hydrogen + Fuel Gas to Heaters	581	582
Hydrogen + Fuel Gas to Export	140	-
PG Ethylene (E)	1,500	1,687
Hydrogen + Fuel Gas to OCM	-	180
Ethane to OCM	-	309
Propane to OCM	-	69
PG Propylene (P)	644	701
1,3 Butadiene (BD)	227	239
Isobutene product	145	150
Raw Pygas	995	1015
PGO+PFO	170	170
Total HV E+P+ BD	2,371	2,627
CO_2 Emission – Heater firing	**1,481**	**1,501**
CO_2 Emission – Export Penalty	**356**	**-**
CO_2 Emission – OCM	**-**	**177**
Specific CO_2 kg/kg ethylene	**1.22**	**0.99**

OCM Integration with a Gas Plant

In a typical gas processing plant, raw gas undergoes several treatment processes to remove acid gas impurities and moisture before being sent to cryogenic section to remove valuable hydrocarbons or NGL. Lean gas is sent back to header as sales gas after compression.

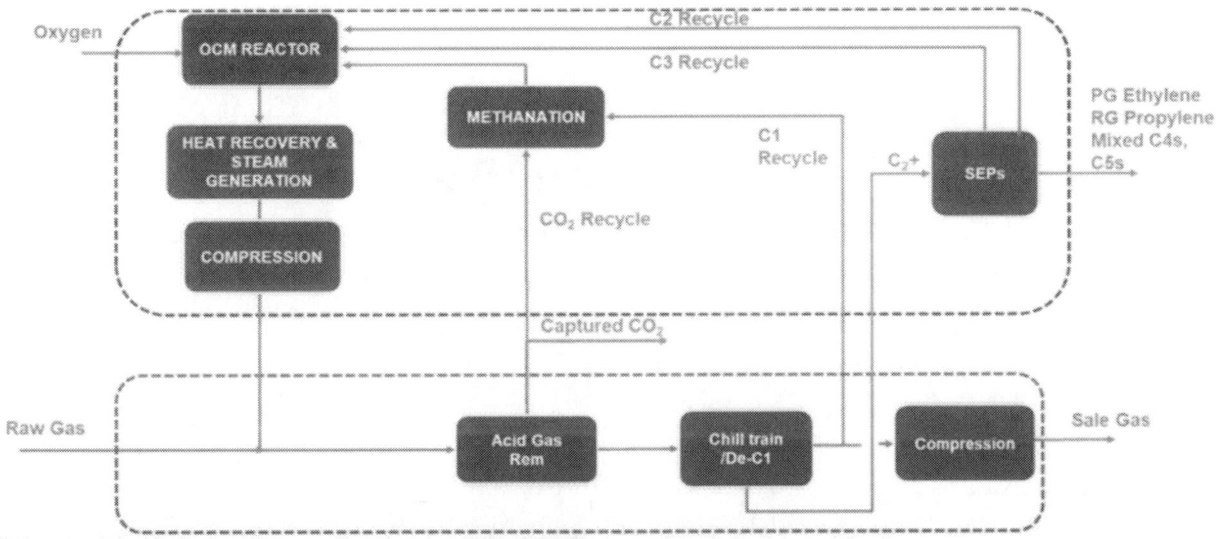

Figure 5: Schematic of a typical OCM/gas plant integration.

When OCM is integrated with a gas plant, several aspects of the OCM flowsheet, shown in Fig-2, can be combined with the existing assets of a gas plant. The integrated flowsheet appears in Fig-5. A split of the sales gas is sent to the methanation section of the OCM. This reactor allows conversion of part of the CO_2 emitted in the process into CH_4 by using a portion of the H_2 generated within the OCM. External H_2 can be added here to recycle all of the CO_2 generated during the OCM process for zero process emissions from OCM. In cases where CO_2 removal is not a priority or where other avenues of CO_2 sequestration are available within the gas plant setup, the methanation reactor can be omitted from the flow scheme. In such cases, the sales gas is sent directly to the OCM reactor along with the ethane/propane recycle from the separation unit.

OCM reactor effluent is used to generate steam and cooled by internal recycles. Cooled OCM reactor effluent is then compressed and mixed with raw gas and sent for further processing into the gas processing system. Demethanizer bottom stream from the gas plant is further processed within the OCM flowsheet to produce polymer grade ethylene and other products. The complexity of the separations system would depend on the desired product distribution and purity. Main benefit of this integration is that existing assets are effectively monetized, thus lowering CAPEX, and providing better margins and attractive payback.

For gas plant-OCM integration, we consider one 850 MMSCFD gas plant as a reference. After recovering 742 kTA NGL, the lean gas or sales gas is 795 MMSCFD.

For this integration, we have considered a 50 kTA OCM unit as an example. To make 50 kTA of ethylene, we see 2.5% reduction in sales gas. Total NGL production increases by 14%.

We have very short payback period because of two reasons:

1. As unreacted methane goes along with sales gas, resulting in no C1 recycle and smaller Methanation unit.
2. Existing gas plant assets can be used.

Table 3: Material balance for OCM integration with gas processing plant.

All values in kTA	Gas Plant	Gas Plant w/ OCM
Raw Gas	6,378 (850 MMSCFD)	6,378 (850 MMSCFD)
Oxygen		146
Lean Gas	5,463 (795 MMSCFD)	5,326 (773 MMSCFD)
Ethylene	-	***50***
Ethane	368	344
Propane	182	184
C_4s	128	130
C_5+	64	69
CO2	173	320
Total NGL	742	653
Water	-	101

Conclusion

There is great interest in the industry to decarbonize the production of base olefins. Lummus' OCM technology offers a path to monetize low value fuel feedstocks into valuable petrochemicals. Integration with existing or grassroots cracker assets provide an excellent opportunity to reduce CO_2 emissions while boosting olefin production.

References

1. Radaelli, Guido, Iyer, Rahul, Rafique, Humera. (2017) "Building with Natural Gas," presented at the 2017 AIChE Spring Meeting, San Antonio, TX
2. Radaelli, Guido, Rafique, Humera. (2019) "Conversion of Low-cost Feedstock to Petrochemical Products," presented at the 2019 AIChE Spring Meeting, New Orleans, LA
3. Radaelli, Guido, Kemper, Rainer. (2015) " Siluria and Linde OCM: Turning A Breakthrough Into Commercial Practice," presented at the 2015 AIChE Spring Meeting, Austin, TX
4. Weinberger , Sam, Scher, Erik, Iyer, Rahul. (2012) " Natural Gas to Ethylene in One Step Siluria Technologies OCM (Oxidative Coupling of Methane)," presented at the 2012 AIChE Spring Meeting, Houston, TX

110b Reducing Carbon Emissions in Ethylene Plants

Myrian Schenk, Andrew Stone, James Middleton

Reducing emissions in ethylene plants is key for its future. Achieving net zero targets by 2050 and limiting increasing global temperature is the challenge: governments and industry need to work together to allow the deployment of a range of technologies that allow achieving the goal. Producing green ethylene is another option, perhaps more distant in the future. Today's technologies allow for reduction in the emissions at a very large percentage.

Options for new and existing ethylene plant include the use of each individually or the combination of technologies such as: electrification, hydrogen firing and carbon capture. >99% of CO2 emissions can be captured.

In this paper we present the possible options and highlight the advantages of combining technologies together with the integration within the cracker:

- Optimal percentage of hydrogen firing vs. hydrogen export

- Selection of electric vs steam drivers to maximise plant electrification

- Carbon Capture unit utilising proven technology (Shell Cansolv)

- Energy optimisation in terms of potential heat recovery from Flue Gas and Heat Integration with the Steam Cracker

- Evaluation and integration of Carbon Capture Unit with existing plant and utility systems.

- Optimisation of cooling load (i.e. air vs cooling water) for the specific plant.

All captured CO2 will meet >99 % purity and be ready for sequestration or utilisation- a number of routes to utilise the CO2 will also be highlighted.

110c Shipping Decarbonization: Energy Assessment of Onboard CO_2 Capture Technologies

Jaafar Ballout, Mamoun Al-Rawashdeh, Dhabia Al-Mohannadi, Joseph Rousseau, Gareth Burton, Patrick Linke

The international shipping industry's greenhouse gas (GHG) emissions are expected to reach 1,500 million tonnes of CO_2 equivalent per annum in the coming years due to the increase in global seaborne trade activities [1]. This would make the shipping industry one of the top 5 GHG emitters, surpassing Japan and ranking below only Russia, India, the United States, and China [2]. Furthermore, the World Economic Forum (WEF) has predicted that GHG emissions from the shipping industry could increase by up to 250% by 2050 if no substantial measures are taken by stakeholders and policymakers [3]. Recently, the International Maritime Organization (IMO) has set new targets aiming to reduce GHG emissions by 20% by 2030, 70% by 2040 compared to 2008 levels, and to achieve net-zero emissions by 2050 [4]. Achieving these goals will require exploring various technologies and solutions to enhance the ships' efficiency and reduce GHG emissions. The shipping industry has already implemented the operational measures, such as optimizing vessel speed [5]. In addition, the adoption of alternative zero-carbon fuels is still lagging behind due to the technical and economic constraints [5]. Consequently, onboard CO_2 capture systems present an attractive option to reduce CO_2 emissions on the ship [6] and meet the IMO's 2030, 2040, and 2050 goals.

The research into onboard CO_2 capture is still in its early stages. Few studies have investigated the feasibility of absorption, membrane separation, and cryogenic technologies for capturing CO_2 onboard and storing it in a liquid form. A comprehensive integration of energy between the ship's main engine and the carbon capture system requires considering the heat and power requirements, which have not been addressed in previous work. Therefore, our study aims to investigate onboard CO_2 capture technologies for an LNG carrier vessel powered by LNG fuel. A comparative energy assessment was conducted to understand the performance of four alternative CO_2 capture technologies: absorption, membrane separation, adsorption, and cryogenic separation. This assessment involved the development of heat and power integration concepts to recover heat from the flue gas stream and meet the heat and power needs of CO_2 capture and compression. Organic Rankine cycles (ORCs) were applied as a heat-to-power conversion technology to fulfill the work requirements in the capture and compression stages. Finally, the liquefaction and storage step was achieved using the LNG fuel stream as a cryogenic source to cool and store the captured CO_2.

In conclusion, the results show that adsorption is one of the promising technologies for onboard applications, achieving a capture rate of approximately 50% using the available energy onboard. This work highlights the importance of thermal-driven CO_2 separation technologies, i.e. absorption and adsorption, in significantly reducing CO_2 emissions from the shipping fleet without the need to burn extra fuel onboard.

Acknowledgements: This work was made possible by joint funding provided by the American Bureau of Shipping (ABS) and Texas A&M University at Qatar.

References:

[1] L. C. Law, E. Mastorakos, and S. Evans, "Estimates of the Decarbonization Potential of Alternative Fuels for Shipping as a Function of Vessel Type, Cargo, and Voyage," *Energies*, vol. 15, no. 20, p. 7468, Oct. 2022, doi: 10.3390/en15207468.

[2] J. Friedrich, M. Ge, A. Pickens, and L. Vigna, "This Interactive Chart Shows Changes in the World's Top 10 Emitters," World Resources Institute. [Online]. Available: https://www.wri.org/insights/interactive-chart-shows-changes-worlds-top-10-emitters

[3] WEF, "New Coalition Aims to Make Zero-Emissions Shipping Plain Sailing," 2019.

[4] DNV, "IMO MEPC 80: Shipping to reach net-zero GHG Emissions by 2050," 2023.

[5] G. Mallouppas and E. Ar. Yfantis, "Decarbonization in Shipping Industry: A Review of Research, Technology Development, and Innovation Proposals," *J. Mar. Sci. Eng.*, vol. 9, no. 4, p. 415, Apr. 2021, doi: 10.3390/jmse9040415.

[6] ABS, "INSIGHTS INTO ONBOARD CARBON CAPTURE," 2022.

25a Improvements in Process Design and Development for Scaling Sustainable Processes

Eloy Flores III, Michael Hartmann, Abraham Romero, Robert T Oslin IV

At SwRI, we engage in the process development and improvement for a variety of sustainable fuels and chemicals from lab to pilot scales. The applied research, development, testing and evaluation methods undertaken help bridge the gap between fundamental science and commercialization. The technologies developed include but are not limited to hydrogen production, CCU, renewable fuels, chemicals and materials production and conventional thermo, photo and electro-chemical conversions. Over the years, immense knowledge has been gained from the design, build-up and operations of such processes at various scales. Many limitations such as feedstock preprocessing and feeding, reactor optimization, downstream processing as well as troubleshooting of pilot scale systems have been overcome and understood. The systems have been developed to ensure process safety, enhanced conversion efficiencies, improved energy and mass transfer as well as overall emissions reductions and carbon neutrality. This session will help transfer this knowledge to the industry and professionals to enable a smooth and sustainable energy transition while bringing fundamental technologies towards commercialization.

25b Commercial Demonstration and Operation of Energy Efficient Olefin-Paraffin Separations with Novel Optiperm^{TM} Membrane System

Brandon Burghardt, Jacob Wargo, Christine Parrish

The separation of light olefins from paraffins is one of the most energy intensive in the world, consuming as much energy as Singapore[1]. The state of the art separation method is distillation, a technology that has remained unchanged over decades. Membranes hold the promise of a more modular, energy efficient and capital efficient method for separation. However, they have historically been hampered by low performance and stability. CMS has developed a transformational membrane technology that is being demonstrated at commercial scale with Braskem.

The first Optiperm™ membrane utilizes a silver embedded polymeric material to selectively transport olefins. This material has demonstrated over 3 years of longevity in laboratory operation and has shown high selectivity and permeance under realistic operating conditions.

In continuing the commercialization of Optiperm™ membranes, CMS' has partnered with Braskem to jointly develop and fabricate a multistage pilot unit to test commercial scale spiral wound membranes in in a mixed C_3 process stream. This talk will discuss results from the pilot unit that was deployed in a Braskem Polymer facility in January 2022 and was brought on stream in May 2022. This is the first commercial scale operation of Optiperm™ membrane modules under field conditions using actual process streams. We will describe the pilot unit design which incorporates all the process equipment of a commercial scale unit including an advanced process control system that enables the unit to simultaneously produce two high quality product streams. Learning from our field testing with Braskem will be shared with effects of the process operating conditions and membrane operating window. Representative plant economics based on field test results will be presented. Based on preliminary calculations, implementation of the technology can affect a 30% decrease in energy use and 70% reduction in GHG as compared to processing the same molecules through the column.

In conjunction with the field unit, CMS developed an internal rig for validation, model validation and testing the extension of the operating window. This talk will discuss the use of the "laboratory twin" for model development. Updates will also be presented on the scale up of our flat sheet spiral wound commercial membrane cartridges, process control systems, and the progress on our joint development project with Braskem.

1. Sholl, D., Lively, R. Seven chemical separations to change the world. *Nature* **532**, 435–437 (2016). https://doi.org/10.1038/532435a

45a Advances in Biorefinery Sustainability Assessment Via Co-Design and Open Platforms

Mark Mba Wright, Pallavi Dubey

Sustainability assessment practitioners have a growing interest to integrate process design, technoeconomic assessment, and lifecycle analysis into cohesive frameworks. These frameworks should enable the rapid assessment of a growing number of biorefinery concept designs and produce lifecycles. Novel computational platforms are enabling the development of these frameworks. However, this development introduces novel challenges in terms of data gathering, model complexity, and result interpretation. Our research group is evaluating ways to advance sustainable process design while increasing design productivity and efficiency.

The BioSTEAM python package is an open source software that enables rapid development and assessment of process design concepts. Our group has developed over a dozen biorefinery designs using BioSTEAM. This presentation describes our experience in the use of BioSTEAM and its impact on designer training and productivity. Furthermore, we discuss ways to integrate machine-based codesign elements to accelerate time to design and result interpretation. Finally, we discuss opportunities for software adoption and integration with various platforms. Specifically, we will discuss the integration into supply chain analysis to understand the design of large-scale adaptive biorefinery deployment scenarios.

45b CFD Modeling in Waste Gas Treatment Processes of Hi-Tech Manufacturing Plants

Minkyu Han, Krishnadash S. Kshetrimayum, Younghee Lee

High-purity inorganic and organic compounds are used either as a reactant or cleaning gas in High-Tech manufacturing plants like semiconductor FABs and electronic display production line, due to which, exhaust waste gas containing both organic and inorganic compounds along with particulate dust are produced which are either toxic or flammable or harmful to environment. Different methods are employed for treating these waste gases depending on the type of pollutants and their required environmental standards for emission. Thermal oxidizers and thermal decomposers are normally used for VOC (volatile organic compounds) and toxic gaseous pollutants, while wet scrubber are normally used in the case of removing acidic or basic gas and particulate dust from the waste gas stream. This presentation will discuss CFD modeling of wet scrubbers and thermal oxidizers for waste gas treatment in high-tech manufacturing plants with a particular focus for design improvement and enhancement of process efficiencies of these equipment whose environmental importance is ever increasing in the midst of today's ESG (Environmental, Social, and Governance) spotlight.

187a Imtex's Permylene Membrane Separation Technology - Transforming Olefin Extraction and Purification for the Petrochemical and Refining Industries

Jamie Hughes

Imtex's Permylene membrane separation technology transforms olefin extraction and purification for the petrochemical and refining industries. Imtex' membrane technology provides a profitable, high-performance, energy efficient alternative to traditional distillation techniques for olefin-paraffin and olefin-nitrogen separations. Permylene seeks to evolve core olefin processes by debottlenecking, decarbonizing and electrifying olefin separation.

Permylene focuses on olefin recovery from polyolefin vent/purge processes, olefin recovery from raffinate streams, and novel propylene upgrading for process debottlenecking, energy savings and greenhouse gas reduction. Imtex Permylene is nearing its goal for full commercializing – in late 2022, Imtex sold its first commercial technology license to a Top 15 Global petrochemical industry player.

Imtex has now run successful onsite customer field trials using plant feed from a polyethylene reactor vent stream and a C4 raffinate, along with extensive C3 separation testing. Imtex is also shortly completing commissioning of a full-scale membrane system at its Customer Demonstration Center where prospective clients can process up to 0.5t/hr of mixed light olefin feed streams through a single module. Imtex expects to start customer application testing in Q1 2024. This paper will present some of the operational and performance data from those trials along with a discussion on the progress and learnings on the path to commercial-scale project implementation.

187c Sustainability Metrics for Small Molecules – Scorecard Focused on Sustainability and Cost

Abraham Schuitman, Dan Valco

This presentation will overview Corteva Agriscience's approach to sustainability metrics at the Discovery and Development stages for small molecules. Quantifiable methods will be presented. The scorecard activities discussed tie directly into our technology development process to drive holistic process improvement. The techniques presented are actively being employed in support of Corteva's 2030 Corporate level sustainability goals. Case studies will also be presented on past route decisions, as well as characterization of these choices/processes in the context of the twelve principles of Green Chemistry. Implementing these strategies has resulted in lower waste, increased safety, and improved cost positions.

70a Multi-Scale Modeling and Simulation of Hydrogen Production Processes

Seckin Karagoz

Because of the broad use of fossil fuels for energy generation, the world is experiencing serious environmental issues. As a result, lowering carbon emissions has become a pressing concern. In this context, hydrogen is receiving worldwide attention because of having huge potential such as becoming the leading energy carrier and its mobile/stationary power generating applications, to be part a solution of the aforementioned problem. A significant reduction on the production cost of hydrogen is required to reach the level of widespread use because this is the key factor for the further development and growth of the hydrogen economy.

Producing clean and cheap power by eliminating the economic and energy penalties requires either the development of new technologies or the improvement of existing energy systems based on sustainability standards and alternative fuels. However, the development of sustainable energy systems, do not negatively influence the economic and societal benefits, is the challenging global target of the energy sector. The incorporation of sustainability into the design of energy systems helps to attain this goal by avoiding or limiting negative impacts. However, monitoring sustainability performance and decision-making are critical and challenging processes for defining sustainability levels. Thus, the development of a methodological approach with proper sustainability indicators ensures the capability of gathering and abstracting the complex process operations and provides straightforward analysis and communication.

To sum up, to eliminate the existing problems and reach the global goals, the research activities need a robust, efficient, reliable methodological approach that spans multi-scale modeling and simulation (MMS)-based analysis, design, synthesis, and optimization. To bridge the gap between atomistic, mesoscopic, and macroscopic systems of interest, powerful computational techniques have been developed. As a result of this development, multi-scale systems engineering studies have produced significant outcomes for the oil, natural gas, and coal industries, as well as the renewable energy and environmental fields.

Thus, the main goal of this work is to identify the role of the multi-scale modeling and simulation (MMS) for hydrogen production processes. The work will consider the following points:

1. Definition of MMS and its principles.
2. Integration of MMS into hydrogen production processes.
3. Implementation of MMS to hydrogen production processes (Current status).
4. Projected status of MMS for hydrogen production processes.

70b Generic Simulant Design Using Statistical Methods and Development

Dylan Rodene, Daniel Odell, Stephen Harris

Savannah River National Laboratory is investigating generic 'base' simulants to blend for upcoming sludge batches of waste product at the Defense Waste Processing Facility (DWPF). A statistical approach has been implemented using two data clustering algorithms: K Means and Normal Mixtures. These data clustering algorithms were applied to a parameter space formed from related past batches processed at the DWPF. Statistical methodology was used to identify the number of general simulants needed to make this task viable. Afterwards the simulants were computationally evaluated against projections to determine goodness of fit and model adequacy. The evaluation was initiated using restricted regression analysis that determined the proportions of simulants needed to match various test cases. The results were corroborated by matching Python and SAS output. Moving forward, concentration will be on optimizing and expanding the design space and model confirmation using regression of the new simulants to target expanded case projections.

70c Utilizing Statistical Doe and Modeling to Accelerate Polyurethane Research

Wenyu Su, Fang Du, Michael Lubitz, Guido Bramante, Jody Henning

The Dow Chemical Company is a world leading material science company where statistical techniques have been widely used to enable efficient scientific research for industrial innovation. Through a real polyurethane (PU) research example, this presentation introduces how we have applied statistical design of experiments (DOE) and modeling to accelerate new product development with statistical confidence.

The DOE study incorporated three categorical and three continuous factors, each with three levels. A full factorial design would require 729 experiments, which is impractical. A 60-run D-optimal design was developed to screen polyol backbones and additives for formulation optimization. The DOE was evaluated to ensure its suitability for prediction and the investigation of factors' impacts, with a focus on aspects such as relative prediction variances, correlation between model terms, and variance inflation factors. The DOE significantly reduced experimental work and still allowed the fitting of relevant statistical models with the DOE data collected.

Based on the DOE, statistical models were developed by integrating various sources of data, such as analytical image analysis, visual inspection, rheology test, and dynamic mechanical test. Several statistical modeling techniques including beta regression, logistic regression, classification tree, and random forest were utilized to determine the impacts of critical factors and identify optimal formulations. Simulation was also performed to assess variable importance in explaining the variability of the performance measurement. The statistical analysis successfully screened the polyol backbones and additives for next-stage formulation optimization. This presentation demonstrates the power of statistical DOE coupled with statistical modeling in accelerating PU research and driving data-based business decisions.

91a Lifetime Value of Simulation: From Point Improvement to Comprehensive Optimization

Ravindra Aglave, Shawn Baggerley, Thomas Eppinger

Chemicals and materials industry is undergoing transition due to two reasons: responding to the energy transition & sustainability requirements and demand for new innovative products to stay competitive.

Designing equipment and processes to satisfy these multiple objectives and constraints by modifying a large number of design parameters is a challenging task since the relationships of these multiple inputs and outputs can be highly non-linear. By analyzing data on multiple variables, engineers can identify the most sensitive factors affecting energy efficiency and performance and select a design or operating condition that meets the process or product requirements. Traditional methods, such as parametric studies can be too time consuming or fraught with computational limitations.

Rigorous optimization algorithms lend themselves well to being applied to solve such multi-input, multi-output problems. An optimization algorithm iteratively runs evaluations and understands the sensitivities between all inputs and combination of inputs to all outputs. The algorithm successively guesses a better design until a converged design is reached that optimizes all the required objectives while satisfying all constraints. This study details the implementation of this optimization methodology to design optimal processes for example cases. We will then introduce how the outcomes can be captured in a way that it can be used throughout the lifetime of the operation.

References

1. Optimization of a divided wall distillation column
2. Optimization of chemical reaction mechanism with HEEDS
3. Optimizing pump designs for process applications
4. CFD based rigorous optimization in the chemical industry

91b A Sustainable Transition Strategy: Electrifying Natural Gas Processing Plants for a Net-Zero Future (A Dual Assessment Approach).

Soha Mousa, Dhabia Al-Mohannadi

A Sustainable Transition Strategy: Electrifying Natural Gas Processing Plants for a Net-Zero Future (A Dual Assessment Approach).

Soha Mousa, Dhabia Al-Mohannadi*

*Dhabia.al-mohannadi@tamu.edu

Chemical Engineering Program, Texas A&M University at Qatar, Doha, Qatar

Industrial processes stand as formidable contributors to global CO2 emissions, accounting for a staggering 37% according to the International Energy Agency (IEA, 2023). The urgency to combat climate change drives the need for transformative solutions. Electrification emerges as a beacon of hope, with the IEA projecting a potential reduction of 2.5 gigatons in global CO2 emissions annually by 2050 through industrial electrification (IEA, 2021). However, challenges loom large, including capital costs and technical intricacies. Electrifying industrial operations, particularly natural gas processing plants, is pivotal in accelerating the transition to a sustainable future. This work investigates the electrification of a natural gas processing plant, employing green electricity generated from renewable sources. By delving into the technical and economic aspects while minimizing environmental impact, this study pioneers a systematic approach towards sustainable industrial practices.

The methodology adopts a comparative approach encompassing three key scenarios: Case 0 represents the existing conventional plant, Case I involves retrofitting and electrifying the conventional plant, and Case II envisions building a greenfield electrified facility. Utilizing advanced energy integration and optimization techniques, we assess these scenarios. Detailed simulations employed provide a robust evaluation of energy efficiency, economic viability, and environmental impact. The preliminary results pave the way for informed decision-making, not only revolutionizing the natural gas processing sector but also setting a precedent for sustainable industrial practices globally.

References

Ahmed, R. O., Al-Mohannadi, D. M., & Linke, P. (2021). Multi-objective resource integration for sustainable industrial clusters. *Journal of Cleaner Production.*

IEA. (2021). *Net Zero by 2050.* IEA.

IEA. (2023, July 11). *IEA.* Retrieved from International Energy Agency: https://www.iea.org/energy-system/industry/chemicals

91c Integrated Multiscale Modeling-Simulation (MMS) and Machine Learning (ML)- Based Design and Development of Novel Systems, and Processes

Seckin Karagoz

Traditionally, the development of novel systems and technologies is complemented by testing. Experimental tools for testing and examining the results are expensive and their use is time consuming. In recent years, thanks to the rapid growth in the computational speed, it is seen that the use of computer aided methods in the systems' design phase contributes greatly to the reduction of the cost and time for the entire technology development process. Experimental methods can improve the computational model by incorporating new data, while computational tools can use this advanced model to analyze structure, properties, and optimization by examining a wide set of possible configurations. Moreover, the simultaneous use of computational and experimental tools allows to tackle problems that cannot be solved using theoretical or experimental methods itself. At this context, overcoming the challenges which hinders commercialization of the technology requires hierarchical theoretical and experimental approaches. Thus, in this session, the focus is the comprehensive understanding and application of Integrated Multiscale Modelling-Simulation (MMS) and Machine Learning (ML) approach which provides new insights for acceleration on the commercialization/utilization for the design and development of novel technologies. The practice of the main concepts is demonstrated through an example of process intensification by reactive-separation systems in power generation process (Integrated Gasification Combined Cycle power plant, intensified by hybrid Membrane Reactor/Adsorptive Reactor system) for simultaneous hydrogen production and CO2 Capture.

57b Effects of Alumina Films on N-Doped Carbon Nanotubes/Graphene Composites As Anode Materials of Lithium-Ion Batteries

Chuen-Chang Lin, Shu-Pei Hsu

A carbon nanotubes/graphene composite is grown on nickel foil without additional catalysts by one-step ambient pressure chemical vapor deposition (CVD). Next, the carbon nanotubes/graphene composite is modified by radio frequency (RF) nitrogen-plasma. Finally, to improve its initial coulombic efficiency/electrochemical stability, lower potential during the charge process (coin cell), and boost potential during the discharge process (lithium-ion battery), alumina is deposited onto the N-doped carbon nanotubes/graphene composite by RF magnetron sputtering at different power levels and periods of time. The charge specific capacity (597 mAh/g) and initial coulombic efficiency (81.44 % › 75.02 % for N-doped carbon nanotubes/graphene) of Al_2O_3/N-doped CNTs/graphene for the coin cell reached a maximum at the best sputtering condition (power = 65 W and time = 30 min). Al_2O_3/N-doped CNTs/graphene (the best sputtering condition) exhibits higher initial coulombic efficiency (79.8 %) compared with N-doped CNTs/graphene (initial coulombic efficiency: 74.3 %) for the lithium-ion battery. Furthermore, the achievement fraction (about 70 %) of full charge capacity (coin cell) for Al_2O_3/N-doped carbon nanotubes/graphene (the best sputtering condition) is higher than that (about 30 %) for N-doped carbon nanotubes/graphene at a voltage lower than about 0.25 V. Moreover, it also shows a little higher electrochemical stability (coin cell) of charge capacity for Al_2O_3/N-doped carbon nanotubes/graphene (the best sputtering condition) in comparison with N-doped carbon nanotubes/graphene and Al_2O_3/N-doped CNTs/graphene (the best sputtering condition) exhibits better cyclic stability (lithium-ion battery) of discharge capacity compared with N-doped CNTs/graphene.

57c Purification of *S*-1-Phenylethanol from the Enantiomer Mixture By Three-Phase Crystallization

Lie-Ding Shiau

1-Phenylethanol is one of the most commonly available chiral alcohols. As each enantiomer has its distinct odor, it is essential to obtain the enantiopure 1-phenylethanol from the racemic mixture for application in food ingredients and in the cosmetics. The enantiopure 1-phenylethanol is also an important precursor in the synthesis of other flavors, fragrances, and pharmaceuticals. Three-phase crystallization (TPC) is a new separation technique which combines melt crystallization and vaporization. Basically, TPC is operated at a triple-point condition, in which the liquid is simultaneously vaporized and crystallized due to the three-phase transformation. In essence, the process is continued until the liquid phase is completely eliminated and only the pure solid crystals remain in the feed. Thus, crystal washing is not required since only pure solid crystals remain in the feed and no impurities are adhered on the crystal surfaces at the end of TPC. In the previous research, TPC has been successfully applied to separate p-xylene from the mixed xylenes by Shiau et al. (2008, 2018). TPC is introduced in this work for chiral purification of *S*-1-phenylethanol from the enantiomer mixture. The operating pressure for the three-phase transformation during the TPC cooling process is determined based on the thermodynamic calculations. A model based on the mass and energy balances is proposed to determine the product yield and enantiomeric purity of *S*-1-phenylethanol for TPC. The experimental yield and purity of the final product are compared with those predicted by the model.

References;

1. Shiau, L. D. "Purification of styrene from a styrene/ethylbenzene mixture by stripping crystallization", *Eng. Chem. Res*, 2018, **57**, p6759-6765.
2. Shiau, L. D., Wen, C. C. and Lin, B. S. "Separation of p-xylene from the multicomponent xylene system by stripping crystallization", AIChE Journal, 2008, 54(1), p337-342.

57d Chevron Richmond Refinery Fire Incident Investigation

Celesta Cox

On August 6, 2012, a pipe rupture occurred in the #4 Crude Unit at a Chevron refinery in Richmond, California. The incident occurred due to thinning pipe walls from sulfidation corrosion on a 52-inch component on the 8-inch 4-sidecut stream exiting the unit's C-1100 Crude Unit Atmospheric Column. This rupture caused the release of light gas oil that was partially vaporized into a vapor cloud that ignited two minutes later. All employees engulfed by the vapor cloud were not seriously injured: eighteen due to escape, and one due to fire-fighting protective equipment. However, this incident caused the release of vapor, smoke, and particulates into the surrounding area. Shortcomings in the following Center for Chemical Process Safety Process Safety Management System (CCPS PSM) pillars were found to be related to the incident: asset integrity and reliability, compliance with standards, emergency management, and safety culture. To the refinery, the CCPS gave recommendations to develop an accountability method to track implementation of best practices, develop an auditable process for appealing denied turnaround work items related to inspection and mechanical integrity, and develop an approval process that included a technical review when lowering the Minimum Alert Thickness of a line. Along with the recommendations given by the CCPS, it is recommended to develop a formal leak response protocol, map refineries for high-risk equipment, and communicate safety survey findings to refineries with action steps.

57e Temperature-Dependent V-Type Isotherm Models: Applied to Water Vapor Adsorption on Metal-Organic Frameworks

Cristian Cardenas

This study offers a comprehensive analysis of the adsorption behavior of water vapor on two distinct metal-organic frameworks (MOFs): aluminum-fumarate and CAU-10. Rigorous determination of equilibrium adsorption isotherms was achieved through a series of controlled experiments encompassing a wide range of temperatures and pressures. The precision of the employed manometric method ensures an exceptional level of data accuracy.

Initially, the Mahle and Weighted Dual-Side-Langmuir models were employed to fit the experimental data and derive isotherm parameters. However, limitations emerged, particularly at low and high pressures, revealing challenges in accurately characterizing water adsorption behavior on the studied MOFs. To address this, a modified Mahle equation was introduced, signifying a substantial refinement of the model, and resulting in a marked improvement in predictive accuracy.

A noteworthy aspect of this study is its dynamic approach. In addition to equilibrium isotherms, the developed model was applied to simulate the dynamic adsorption of water vapor within a fixed-bed column. This dynamic simulation affords a more comprehensive understanding of how water vapor interacts with MOFs over time, providing invaluable insight into the transient behavior of the system.

The implications of this research extend far beyond the laboratory. The findings hold great promise for practical applications, particularly in the realm of heat pumps and related technologies. Understanding how MOFs interact with water vapor under varying conditions opens avenues for leveraging their unique properties in applications related to moisture capture and management.

Furthermore, this study contributes substantially to the growing body of knowledge surrounding MOFs, a class of materials that has garnered considerable attention for their potential in various industrial applications. The distinctive structure and tunable properties of MOFs make them promising candidates for a wide array of applications, including storage, separation, and gas catalysis.

The outcomes of this study mark a significant advancement in the understanding of water vapor adsorption on metal-organic frameworks. Through a systematic analysis of equilibrium adsorption isotherms and the development of a modified Mahle equation, the research team has made substantial progress in accurately modeling this intricate process. Dynamic simulation of water vapor adsorption in a fixed-bed column further amplifies the practical relevance of these findings.

The results of this study are poised to influence a myriad of fields, from materials science to environmental engineering. The potential applications of MOFs in heat pumps and related technologies represent a tangible pathway towards more efficient and sustainable moisture management. This research serves as a testament to the ongoing quest for innovative solutions in the field of adsorption science and materials engineering.

57f N.D.K. Crystal Inc. Explosion with Offsite Fatality

Olivia Dohm

On December 7, 2009, Vessel No. 2 at NDK Crystal Inc. facility in Belvidere, Illinois erupted, causing a white cloud of steam to be released into the air. The force of the eruption broke the base of the steel vessel and led to shrapnel flying offsite. One fragment of steel flew 435 ft. away from the manufacturing area and hit the wall of Grupo Antolin, an automotive supply company close by. One Grupo Antolin employee was injured by the blow. Another fragment of steel flew 650 ft. away and hit a truck driver who had stopped at the Oasis gas station near I-90 highway. This driver later passed away. NDK Crystal, Grupo Antolin, and cars in neighboring parking lots sustained varying degrees of damage. NDK Crystal was later fined for disregarding various national standards. U.S. Chemical Safety Board (CSB) investigators were sent on the scene to discover what caused the incident. After conducting non-destructive evaluations and destructive testing methods on fragments of the autoclave, CSB made many recommendations. CSB encouraged NDK to be prohibited from using the autoclaves until Illinois deemed them safe for use. CSB also recommended NDK find inherently safer ways to produce crystals, like at lower temperatures and pressures. CSB advised The American Society of Mechanical Engineers (ASME) to update their standards on vessel thickness, as NDK had vessels that were susceptible to corrosion due to being overly thick. Along with this, several other recommendations were necessary. This research evaluates the recommendations made by the CSB and assesses the sequence of events that caused this incident.

57h Synthesis and Evaluation of a Diels-Alder Polyphenylene with Tethered Multication Moieties

Ana Soares, Michael Lee, Nolan Hendrickson, Chris Cornelius

With the increase in energy consumption worldwide, the need for cost-effective and efficient energy conversion and storage systems escalates. Many of the potential robust and scalable technologies require enhancements in ion transport which can be done with enhanced electrolyte materials. Organic membranes with ionic functionality, or ionomers, have been investigated as an inexpensive electrolyte however there have been issues with their usage. Nafion, a common ionomer used, has experienced degradation of the polymer in acidic and oxidative environments, ion crossover that leads to self-discharge, and high scale-up costs. More resistant and price-efficient membranes have been investigated as an alternative. Diels alder poly(phenylene) membranes with methyl groups on the phenyl rings have been brominated and substituted with quaternary ammonium to create benzyl trimethylammonium groups (QAPP) as an alternative. QAPP membranes have been shown to exceed Nafion in terms of selectivity, mechanical robustness, ionic conductivity, and thermal stability. However,
QAPP has been shown to oxidize earlier than Nafion in ex-situ degradation studies. The purpose of this
study is to investigate homogenous functionalization with potentially more robust multication pendant groups. The overall pathway to produce the anionic exchange membrane will involve a novel step in the
monomer precursor synthesis. Glass transition and decomposition temperature will be measured with thermal gravimetric analysis and modulated differential scanning calorimetry. Additionally, in-plane conductivity will be calculated after running impedance test at temperatures ranging from 30C to 80C and water saturation ranging from 10% to 100% and compared to Nafion standard data collected under
the same conditions. Future studies include further transport property analysis such as experimentally determined diffusion coefficients for water vapor, proton conductivity, and liquid water. Additionally, application testing is planned inside flow batteries and a single fuel cell. Also, cell diffusional studies will
show the ionomer's potential to prevent crossover through Donnan effect. Finally, using Fourier transform infrared spectroscopy under attenuated total reflectance at various incident angles, transmission electron microscopy, and grazing-incidence small-angle x-ray scattering will give insight on
the structure layout and chemical composition of the membrane.

57i Unraveling the Impact of Aging on Diels Alder Polyphenylene (DAPP) for Applications below Tg

Nolan Hendrickson, Ana Soares, Michael Lee, Chris Cornelius

Aging, a process occurring below the glass transition temperature (Tg), significantly influences the mechanical viscoelastic behavior of polymers. Extensive research has focused on characterizing the relaxation mechanisms during aging. While numerous studies have investigated bulk thermodynamic properties and microstructural patterns, disparities persist. The implications of these aging phenomena span multiple industries. Notably, high Tg polymers remain inadequately examined. This study centers on Diels Alder Polyphenylene (DAPP), characterized by its high strength and molecular weight, with the objective of bridging the knowledge gap regarding aging's impact on Polyphenylene (PP) and its suitability for industrial applications below Tg. A novel research approach explores how cooling rate and temperature affect PP's aging properties, specifically targeting the temperature range up to 350°C, where PP materials retain their processability. Our investigation employs a diverse range of characterization techniques. We utilize gas permeation measurements to assess transport properties, density measurements to gain insights into chain relaxation analysis, and modulated differential scanning calorimetry (DSC) to link cooling rate effects with observed Tg changes and enthalpic recovery. Additionally, birefringence measurements track alterations in stress patterns. We anticipate that our study will offer valuable insights into the evolution of gas transport properties over time during aging, encompassing reduced permeability and increased selectivity rates. Intriguingly, preliminary findings indicate a decrease in density rather than an increase. These results will contribute to a deeper understanding of DAPP's aging behavior, providing essential data for optimizing its industrial applications below the glass transition temperature. Future studies will expand our knowledge base by obtaining additional microstructural evidence through techniques such as transmission electron microscopy and investigating the mechanical properties within the linear viscoelastic region, encompassing creep compliance, stress-relaxation modulus, and other dynamic mechanical properties like stress-strain.

57j An Examination of the Toxic Chemical Release at Mgpi Processing, Inc.

Olivia Clark

On October 21, 2016, a toxic plume emerged from the sodium hypochlorite tank in the Mod B facility of MGPI Processing, Inc. Located in Atchison, Kansas, the plant manufactured wheat starches, proteins, and distilled products. The toxic release occurred after a miscommunication between an MGPI operator and a delivery driver from Harcros Chemicals, resulting in the inadvertent mixing of sodium hypochlorite and sulfuric acid. The two incompatible chemicals reacted and formed a plume that contained chlorine gas. Many Atchison residents were ordered to shelter-in-place or evacuate, and over 140 people sought medical attention in response to the plume. The Chemical Safety Board (CSB) investigated the accident and published an official report detailing the release. They found a lack of many Risk-Based Process Safety elements including compliance of standards, safety competency, asset integrity and reliability, training and performance, conduct of operations, management of change, emergency management, and process safety culture, all of which were root causes of the incident. The CSB recommended MGPI update their ventilation and add additional controls, Harcros Chemicals provide more training to their delivery drivers and determine if respiratory protection is needed, and the Atchison County Department of Emergency Management study the local chemical facilities and train to respond to chemical emergencies. Further recommendations should be made to MGPI, including the performance of regular equipment checks and the incorporation of labels and remotely operated valves to the chemical fill lines. Moreover, MGPI should provide additional training to operators on chemical deliveries, provide additional PPE for such deliveries, and audit operators on their procedural compliance.

57k Comparative Evaluation of Surfactants with Diverse Tail Structures for Clay Swelling Inhibition

Muhammad Shahzad Kamal, Syed Muhammad Shakil Hussain, Mobeen Murtaza, Shirish Patil, Mohammad Alotaibi, Mohanad Fahmi

Clay swelling poses a significant challenge in the oil and gas industry, especially during drilling and production operations. These clays, prevalent in geological formations, have a tendency to expand, leading to wellbore instability, potential drilling complications, wellbore instability, and diminished production efficiency. Surfactants have emerged as a promising solution to mitigate clay swelling. These chemicals are required to demonstrate solubility across diverse water compositions, endure reservoir ion levels, and remain stable at high temperatures. In this study, we evaluate the efficacy of different surfactant categories, including those with hydrocarbon-based alkyl tails and surfactant having magnetic properties. Our in-house synthesized surfactants exhibit exceptional solubility across varying water compositions and salinity levels, with no evidence of precipitation or phase separation. Moreover, these surfactants maintain their structural integrity, even after prolonged exposure to elevated temperatures (90°C for several days). The swelling inhibition characteristics of our developed surfactants outperforms commercial surfactants. Experimental data indicates that developed surfactants provide prolonged clay swelling inhibition compared to traditional inhibitors like KCl. Furthermore, developed surfactants exhibit improved filtration and rheological properties when compared to commercial alternatives. The diverse range of in-house-produced surfactants showcases their immense potential for effective clay swelling inhibition in the oil and gas industry.

57l Philadelphia Energy Solutions (PES) Refinery Fire and Explosions

Alejandro Lira

This presentation provides a comprehensive analysis of the critical incident that occurred at the Philadelphia Energy Solutions (PES) hydrofluoric acid (HF) alkylation unit in June 2019, emphasizing the importance of safety in the management of hazardous materials. The incident involved a rupture of a steel elbow pipe, leading to a catastrophic chain of events, including a large vapor cloud release and subsequent fire, followed by a series of explosions. The investigations revealed that several factors contributed to this incident. Notably, a lack of inspection requirements for carbon steel piping in HF corrosion zones, along with equipment specification issues. Additionally, the limitations of certain 'active' safeguards were highlighted, and it was shown how they can be compromised in fire and explosion scenarios.

The presentation also outlines five key safety issues identified in the PES alkylation unit incident, including concerns related to mechanical integrity, equipment verification, emergency isolation, safeguard reliability, and the application of inherently safer design strategies. To address these issues and enhance safety in HF alkylation units across the industry, the Chemical Safety Board (CSB) put forth a series of recommendation. These recommendations include developing an EPA program that prioritizes inspections of HF alkylation units and revising the EPA Risk Management Plan (RMP) to require Safer Technology and Alternative Analysis (STTA) for both new and existing HF alkylation units. Furthermore, the CSB urges a revision of the 5th edition of API RP 751 to bolster safeguard protection from fire and explosion hazards, with the addition of including the installation of emergency isolation valves. Finally, there is a call for ASTM International to revise ASTM A234 and incorporate supplementary requirements for piping used in HF service.

There is also an emphasis on the critical need for the proactive safety measures, inspection protocols, and the adaptation of inherently safer technologies in the handling of HF. By drawing lessons from this incident and embracing these recommendations, it's possible to work towards a safer and more sustainable future in the production of hydrocarbon products.

57m Optimization of the Use of Pyrolysis Waste Oil As a Feedstock for NCC Process through Hydrotreating and Hydrocracking

Sukhyun Lim, Eun-Hee Kwon, Hung Pham, Kwang Ho Kim, Young-Hwan Chu, Kang-Seok Go, Nam Sun Nho

This study was conducted to upgrade waste plastic pyrolysis oil (WPPO) produced from a 10 ton/day commercial rotary kiln waste plastic pyrolysis reactor as a feedstock for the NCC process. To achieve this objective, hydrotreating and hydrocracking reactions were carried out utilizing both laboratory and bench-scale continuous process equipment. The research mainly aimed at designing the process concept for the continuous unit in the laboratory on a small scale. The research findings confirm the impact of various combinations of HDT and HCK reactions, as well as their sensitivity to feedstock characteristics and reaction conditions. In conclusion, the impurity removal rate met the majority of naphtha specifications, and the naphtha yield exceeded 90%.

57n Simulation of Bitumen Partial Upgrading Process

Hung Pham, Nam Sun Nho, Kwang-ho Kim, Youngdoo Kim, Sun-A Choi, Eun-Hee Kwon, Sukhyun Lim, Young-Hwan Chu

Bitumen partial upgrading is a process that transforms bitumen to reduce its diluent requirements, thereby lowering transportation costs and increasing profit. However, this process has failed to meet the product quality specifications for partial upgrading (density <0.9400 g/cm3 at 15.6°C and viscosity <350 cSt at 40°C). Therefore, there is a need to develop an improved partial upgrading process. In this study, we constructed a plant-wide process model in Aspen HYSYS that incorporates visbreaking (VB) and solvent deasphalting (SDA) based on experimental data from batch and bench-scale experiments using VR Feed. Then, simulation studies were performed to evaluate the process configuration by changing the sequence of VB and SDA and using Wabasca Primary feed. The comparison between configurations was evaluated in terms of product yields, diluent requirement, operating cost, and capital cost. Finally, we suggested an optimal configuration that balances all goals.

57o Unleashing the Power of Hybrid Modeling for Process Design and Optimization

Reza Haghpanah, Brad Metzler, Evan Bergman, Aayush Singh, David Vickery

Process optimization via mathematical modeling is a powerful tool in the chemical process industry for developing and implementing process improvements. For example, mathematical models can be used to simulate different process scenarios and identify the optimum operating conditions that will maximize productivity while minimizing energy usage. This can lead to significant energy and cost savings, increased capacity, and profitability for the industry. However, developing accurate first principles (physics-based) models can be complex, especially when modeling large and complex processes with multiple physical and chemical phenomena. Hybrid models, which combine physics-based models with machine-learning models or data-driven models, can be used to address some of these limitations. By combining different modeling techniques, hybrid models can provide a more accurate and comprehensive representation of the process while reducing the computational cost of the model. This work presents the use of Aspen Hybrid Models™ at Dow in a highly integrated process flowsheet. First-principles-driven hybrid models were developed in Aspen Plus V12.1 for reactors with complex reaction networks coupled with distillation columns and multiple recycle streams. Methodology for development of the hybrid model will be shared along with high level results of model validation versus plant performance and model optimization opportunities.

57q New Viscosity Models Based on Machine Learning and Classical Regression

Adel Elsharkawy, Osamah Alomair, Hitham Lababidi, Asrar Alshammri

Several research publications and technical papers have been dedicated to the modeling of crude oil viscosity. While experimental measurements are the most reliable means of determining oil viscosity, they are costly, time-consuming, and sometimes unfeasible due to sample unavailability. To address such challenges, various approaches like correlations, machine learning, artificial neural network models, and data mining methods have been utilized to estimate crude oil viscosity. Nevertheless, some of the published models suffer from inherent limitations. These limitations arise from their reliance on limited data and their tendency to focus on specific types of crude oil, which ultimately leads to reduced accuracy in viscosity predictions. This research aims to overcome these limitations by employing experimentally measured crude oil viscosity data from 66 samples collected at different pressures and temperatures from the Burgan oil field in Kuwait. The study presents multiple methods for predicting the viscosity of dead, saturated, at bubble point pressure, and undersaturated oil.

The most well-known and published models have been used to assess their accuracy in predicting the viscosity data of the 66 samples described in the data bank. It was found that Labedi (1992) model is the most accurate in prediction of the dead oil viscosity with an average absolute relative error (AARE) of 13%. Katoatmodjo et al. (1991) was found to have the best accuracy among all the published correlation for saturated crudes with an AARE of 8%. Abu Khamsin et al (1991) has provide the best accuracy for oil viscosity at bubble point pressure with an AARE of 14% and Abdulmajeed et al (1991) has the best accuracy for undersaturated crude with an AARE of 5.8%.

A 990 measurements has been used to developed machine learning models (MLM) and classical regression models (CRM) for the predictions of dead oil viscosity at atmospheric conditions saturated oil viscosity at pressures below the bubble point pressure, at bubble point pressure and above the bubble point pressure. The proposed CRM was able to predict the dead oil viscosity with AARE of 6.1%, the saturated oil viscosity below bubble point pressure with AARE of 5.2%, the oil viscosity at bubble point pressure with AARE of 6.3%, and the undersaturated oil viscosity above bubble point with AARE of 2.2%.

149a Risk Mitigation in Process Scale up of Aminopropyl Silicone Intermediates

Michael Telgenhoff, Brian Rekken, Xiaoyun (Shawn) Chen

Within silicones, access to terminal aminopropyl silicone polymers has been limited due to the difficulties in synthetic methods of aminopropyl silicone endblocking intermediates. Bis-aminopropyl disiloxane is generally used to build terminal primary aminosiloxanes via a silicone equilibration process. However, this intermediate requires multiple synthetic steps and raw materials making production costly. In addition, the production chemistry requires based-catalyzed equilibration with dimethyl silicone followed by neutralization and filtration. An alternative synthetic method involves the primary amino, 2,2,4-trimethyl-1- aza- 2- silacyclopentane (CAS: 661492-94-0) or N-[(3-Aminopropyl)dimethylsilyl]-2,2-dimethyl-1-aza-2-silacyclopentane (CAS: 388606-32-4) as a direct reaction with silanol terminated silicones. However, synthetic yields have been low or a catalyst was required for the silanol capping reaction. Another route that was not previously investigated involves the use of cyclic-siloxazanes, specifically 2, 2, 7, 7- tetramethyl-1- oxa- 3- aza- 2, 7-disilacycloheptane. The cyclic siloxazane readily ring opens at the Si-N bond when combined with silanols to add a disiloxy-propylamine unit. This will occur at ambient temperature and free of catalyst. Prior to the work reported in this paper, the process to make the cyclic-siloxazane (GE Silicones) was plagued by low-volume efficiency and a hazardous, untenable process in a manufacturing setting. Attempts by GE to reduce excessive use of allylamine led to disastrous results —a runaway reaction. As it will be shown in this presentation, the large exotherm was likely a result of an accumulation of SiH and olefin functionalities in the mixture concurrent with the reduction of free allylamine, which is known to act as a hydrosilylation inhibitor. In this work, we show that the cyclic siloxazane can be successfully and safely synthesized through new process controlled methods. The scale up from a glove box to a 1L fume hood reactor with batch distillation was successfully demonstrated for the production of cyclic siloxazane endblocker. There are four key technologies specific to the hydrosilylation of olefinic amines in this example that are required and enable the sensitive chemistry. These include pulsed feed of tetramethyldisiloxane, use of a hydrosilylation promoter, ammonium sulfate catalyzed rearrangement, and batch distillation of the lights and product. Proper implementation of these methods can result in volume yields ~89% at >99% purity.

149b The Most Common Pilot Plant and Laboratory Unit Commissioning Problems

Richard Palluzi

Commissioning, the start up of a newly constructed piece of equipment, is always a challenging problem. It is even more challenging in a research environment as the operation is more novel, the process information and operation less defined, and the equipment performance often less certain. Compared to a plant start up, the commissioning of a pilot plant or laboratory unit generally is less well organized and certainly less fully staffed.

Based on my experience here are the most common problems I see again and again. Suggestions to avoid them are included.

149c Chlorinating Agents Cause Unexpectedly High Corrosion in the Presence of Acetonitrile.

Yamini Krishnan, Jayachandran Devaraj, Ajit Mishra, Paul Speakman, John Doyle III, Abraham Schuitman

Sulfuryl chloride (SO_2Cl_2) is a liquid reagent that that can be used as an alternative to chlorine gas in chlorination reactions. Chlorination reactions with SO_2Cl_2 are anhydrous due to its high reactivity with water. SO_2Cl_2 diluted in dichloromethane (DCM) was used in a chlorination of an aryl molecule as part of the process development for a specialty chemical intermediate . In order to optimize the reaction kinetics, the effects of temperature, pressure, and solvent were studied in different reactor types. Acetonitrile was used to enhance the solubility of the starting material. Over the course of these studies, severe corrosion was observed in the Hastelloy C276 reactor when the reaction was conducted with SO_2Cl_2 diluted in a mixture of DCM and acetonitrile (MeCN). Interestingly, the corrosion was negligible when acetonitrile was eliminated from the reaction mixture. Following these findings, a systematic coupon study was conducted without the reaction reagents (i.e., with mixtures of chlorinating agent + DCM + MeCN or chlorinating agent + DCM alone) at two different temperatures using eight alloys and five chlorinating agents. The coupon test confirmed the results from the initial experiments, with unexpectedly high corrosion rates (>100 mils per year) being observed across multiple alloys whenever MeCN was included in the mixture. This presentation will provide an overview of our results and discuss the implications for reagent selection and process safety.

167b Simulation of a Gas-Generating System Under Runaway Conditions: Studying the Behavior during Venting and Assessing Current Approaches for Evaluating the Gas Generation Rate

Amani Grati, Luc Vechot, Marcelo Castier

Emergency relief systems (ERS) serve as the last line of defense against the explosion of a vessel enclosing a reactive gas-generating mixture under runaway conditions. ERS design for gas-generating systems requires the evaluation of the gas generation rate under runaway conditions. Current methodologies, mainly developed through the work of the Design Institute for Emergency Relief Systems, are based on the experimental evaluation of the gas generation rate under runaway conditions using pseudo-adiabatic calorimetry at a laboratory scale. Such experiments involve the measurement of the temperature and pressure resulting from the runaway in a test cell in either a closed or open (to a containment vessel) cell configuration. The ideal gas law is used with the measured gas temperature and pressure to assess the gas generation rate. Although this methodology is relatively easy to implement, there is currently no consensus on the best experimental conditions, cell configuration, and their interpretation (correction for the effect of thermal inertia) to measure the gas generation rate.

In order to improve the methods used to evaluate the gas generation rate and, to a greater extent, the ERS design methodologies, it is crucial to gain a comprehensive understanding of the phenomena that occur in a vessel containing a gas-generating mixture under runaway conditions before and during the venting process. As a step in this direction, the paper will discuss the use of a dynamic simulator capable of rigorously calculating the temperature, pressure, and phase composition in a vessel containing a gas-generating reactive mixture under runaway conditions. A series of simulations of runaway reactions under venting conditions was performed. The results are analyzed to highlight the behavior of the gas-generating reactive system before and during venting, with a special emphasis on the temperature, pressure, and phase composition at each time step of the runaway and how they are affected by the ERS diameter and the ERS set pressure. The gas generation rate rigorously calculated by the model is then compared with the gas generation rate calculated using the above-mentioned methodologies based on pseudo-adiabatic calorimetry, with an emphasis on the closed-cell test approach.

167c Online Gas Line Purging System for Removal of Explosive Gases

Abhineet Raj, Naveen Kumar

Problem Statement: - Insufficient purging of BF gas line due to manual interventions.

Challenge:

100% BF gas line purging before and after job completion.

Innovation:

Remote purging activity controlled from DCS/Control room

Safety Precautions:
1. Nitrogen pressure should be minimum 6 bar
2. Communication should be clear to all working agency.

Result/Benefits:

Continuous monitoring of CO and O2 concentration.

Avoidance of manual measurement of at gas concentration site to ensue sufficient (100%) purging.

132a Parametric Analysis of Membrane Properties on Post Combustion Flue Gas System Economics

Joshua Watson, Christine Parrish

Many different methods are required to address global decarbonization. While the transportation sector can rely on electrification and power generation can utilize renewable sources, the path to industrial decarbonization requires a more innovative approach[1]. Industrial applications are more likely to need high heat sources and have long lived assets in the ground that are key to producing the chemical building blocks of our society. These applications, like cement, steel, and refractory materials, need low cost carbon capture systems to enable their decarbonization goals. Membranes have a unique modular nature and offer simplified, low cost, low energy systems that are a lynchpins for industrial decarbonization. When applied to post combustion flue gas, membranes can address hard to abate sectors like cement and steel[2] which are responsible for 15% of the global CO2 emissions.

This talk will discuss the carbon capture system design consideration for post combustion flue gas applications. The capital and operational costs are driven not only by the membrane sizing and costing, but also by the compressor costing and electricity use. System design optimization is paramount to determining best membrane staging design and balance of equipment[3]. To best understand this, Compact Membrane Systems(CMS) developed a model to investigate optimal staging designs. This talk will discuss the novel superstructure modelling approach used to enable membrane staging optimization for given membrane performances to find the lowest overall cost. To understand the tradeoffs, a parametric analysis was performed with membrane properties between 100-5000 GPU and 10-300 selectivity. Traditionally, higher selectivity allows for fewer stages to reach high purity products, and increased permeance, decreases the amount of membrane area needed for a given flow rate. Both are important to lowering the capital costs associated with the full system.

According to CMS's analysis of compression cost, high performance at low pressure operation is a necessity for economic carbon capture (CC) with membrane systems. In the hard-to-abate CC applications studied, it was found that above a threshold selectivity, increased permeance unlocks significantly lower costs of capture but increased selectivity yields diminishing economic returns. Realistic capture costs were calculated with amortized installed capital estimates and industry relevant utility, maintenance, and labor cost estimates. Previous research has shown that membranes with low cost of manufacturing can be used with low pressure designs to provide low costs of capture[4,5]. Because membranes with increased properties can be more expensive to manufacture, the implications of membrane costing were analyzed to determine the economic benefits of permeance and selectivity. According to these results, a high permeance membrane with modest selectivity (*i.e.*, 1000 GPU and 20 CO_2/N_2 selectivity) can achieve capture costs under \$50/ton with conservative costing assumptions. A high selectivity membrane with modest permeance (*i.e.*, 200 GPU and 100 CO_2/N_2 selectivity) cannot achieve capture costs under \$90/ton with the same costing assumptions.

1. Larsen, J. (2022, August 18). "A turning point for US climate progress: Assessing the climate and clean energy provisions in the Inflation Reduction Act." *Rhodium Group*. Retrieved April 14, 2023, from https://rhg.com/research/climate-clean-energy-inflation-reduction-act/
2. Baker, Richard W., et al. 'CO2 capture from cement plants and steel mills using membranes.' *Industrial & Engineering Chemistry Research*57.47 (2018): 15963-15970.
3. Arias, Ana, et al. "Optimization of multi-stage membrane systems for CO capture from flue

gas." *International Journal of Greenhouse Gas Control.* 53 (2016). 371-390.

4. Han, Yang, Yutong Yang, and WS Winston Ho. 'Recent progress in the engineering of polymeric membranes for CO2 capture from flue gas.' *Membranes*10.11 (2020): 365.
5. Merkel, Tim, et al. *Pilot testing of a membrane system for postcombustion CO2 capture.* Membrane Technology And Research, Incorporated, Newark, CA (United States), 2015.

132b Application of Machine Learning in Developing Membrane Technologies

Seckin Karagoz

Membrane-based technologies are advancing rapidly. They are used in a variety of critical applications such as gas separation (carbon dioxide separation, air separation, hydrogen recovery, and many more) and water treatment. The materials utilized in these systems, however, determine their performance. Membrane technology advancements, powered by recent advances, are currently capturing more data than ever before. However, the synthesis and development of next-generation materials takes time and money. To enhance membrane technology, data processing and interpretation must be both time and cost effective. Various theoretical methods and models have been developed to speed material discovery. Because of their assumptions and hypotheses, these models are unable to characterize complex materials and systems. The recent rise of machine learning as a powerful tool for combining multimodality and multifidelity data and finding correlations between interrelated phenomena has a unique promise in this field for tackling these difficulties. Thus, the objective of this work is to develop hybrid machine learning models to estimate membrane performance, including permeability, selectivity parameters, etc., for CO_2 capture.

6a Emerging Process Development Areas for Converting Sustainable Feedstocks into Low Carbon Transportation Fuels

Shyamal Bej

For accelerating decarbonization in the transport sector, there is an urgent need for transitioning the currently used largely fossil-based energy mixture into one which should be predominantly of sustainable origin. Appropriate processes need to be developed for converting various future sustainable feedstocks such as biomass and captured CO_2 to produce different types of low carbon transportation fuels. Knowledge gained during earlier transitions as well as novel innovations will be required for developing these processes. After providing an overview on the status of the technologies required, this talk will present examples where significant commercialization success has been achieved leveraging earlier knowledge. The presentation will then focus on a couple of selected cases where further innovations and improvements will be required for their successful deployments. Directional guidelines for bringing these improvements will also be covered in this talk.

AUTHOR INDEX